Office ergonomics

Preventing Repetitive Motion Injuries & Carpal Tunnel Syndrome

By Susan Orr, M.A.
Edited by Howard Allan VanEs, M.A.

Published by

Letsdoyoga.com

Howard VanEs, Publisher

3360 McGraw Lane

Lafayette CA 94549

415-309-1290

www.letsdoyoga.com

info@letsdoyoga.com

Copyright 2007 by Howard Allan VanEs. All rights reserved. Except as permitted under the United States Copyright Act of 1976, no part of this publication may be reproduced or distributed in any form or by any means without prior written consent of the publisher.

Disclaimer

The information in this book is not intended to diagnose or treat any health condition and is for educational purposes. If you are experiencing any pain, numbness, stiffness, weakness or any other symptom in any part of your body it is highly advisable to seek the advice of a physician or qualified health care practitioner.

Acknowledgements

Letsdoyoga.com would like to acknowledge and thank the following people whose contribution was integral to the production of this book:

Author: *Susan Orr (arloine@comcast.net)*

Designer: *Dare Porter (www.realtimedesign.com)*

Cover designer: *Howard Petlack (petlack@hotmail.com)*

Free Wellness Newsletter

As a service to our readers Letsdoyoga.com publishes a wellness newsletter. This monthly newsletter features articles and ideas to help you live a happier and healthier life as well as insights, tips and ideas to deepen the practice of yoga. To receive a copy of Yoga Health and Wellness Newsletter visit **www.letsdoyoga.com** and enter your email address.

Letsdoyoga.com

Letsdoyoga.com offers seminars and workshops on the following wellness related topics:

- Stress management
- Office ergonomics
- Insomnia
- Anxiety
- Tight shoulders
- Back care
- Abdominal exercises
- Meditation
- Pranayama (yogic breathing)
- Secrets of stretching
- Various Yoga oriented workshops

To have any of the above trainings presented to your organization, please contact Howard VanEs at **info@letsdoyoga.com** or **415-309-1290**.

Table of Contents

- About the Author 9
- Introduction 11
- Repetitive Motion Injuries/Carpal Tunnel Syndrome Defined 13
- Impact of Repetitive Motion Injuries 19
- What is Ergonomics and How Can It Help? 21
- Treatment and Causes 25
- Prevention 31
  - *Computer Workstation Set-up and Posture* 31
  - *Working Surface Height and Tilt* 34
  - *Centering* 35
  - *Chairs* 39
  - *Sit/Stand Stations* 44
  - *Footstools* 45
  - *Keyboards* 46
  - *Mice* 50
  - *Monitors* 55

- – Vision/Eye Strain 57
- – Awkward Postures 60
- – Laptops 62
- ▸ Fixes that Often Don't Work 65
- ▸ Behavioral Prevention Tips 69
- ▸ Websites 73
- ▸ OSHA Regulations 75
- ▸ Associations & Publications 77
- ▸ Ergonomic Computer Equipment and Accessories . 79
- ▸ **Bonus Section**: Exercises to Relieve Stress at Computer and Workstations 83
  - – Belly Breathing 83
  - – Eyes 85
  - – Neck 85
  - – Shoulders 86
  - – Wrists, Hands, and Fingers 87
  - – Back and Chest 88
  - – Legs and Hips 89
  - – Rest 90
- ▸ Also Available from Letsdoyoga.com 91
  - – Tight Shoulder Relief 93
  - – Beginning Yoga: A Practice Manual 95
  - – Shavasna/Deep Relaxation (Audio CD) 97
  - – Office Ergonomics: Preventing Repetitive Stress Injuries and Carpal Tunnel Syndrome 99

About the Author

Susan Orr is a Certified Safety Professional (CSP), a certified Infrared Thermographer, and certified in Occupational Hearing Conservation. She holds a Master's degree in Industrial Safety from USC with thesis work in low-level industrial noise and short-term memory task error rates. She has an Associate in Risk Management (ARM) designation and is a member of the American Society of Safety Engineers.

Currently, Susan is a Risk Control Specialist in the Global Technology Department for St. Paul Travelers Insurance working with high tech companies in the fields of biotechnology, computer chip/hardware manufacturers, software/information technology and telecom. The focus of her work includes analysis and evaluation of organizational employee safety and health programs, workplace safety inspections, safety training, fleet risk management, evaluation of fire prevention, disaster preparation/business continuity and security efforts. Susan also serves on the St. Paul Travelers risk control committee monitoring the nanotechnology industry, focusing on employee safety and health.

Susan Orr can be reached via email at: **arloine@comcast.net**

Introduction

Did you know that 90% of the workforce in the U.S. uses a computer? And did you know that the more you use a computer, the more likely it is that you will have a repetitive stress injury? In fact, the risk of a muscle-skeletal injury for someone who works on a computer four or more hours per day is nine times greater than it is for a person who spends just one hour per day on the computer. With so many people using computers and at high risk for injury, it is no wonder there are so many repetitive stress injuries!

> ... the risk of a muscle-skeletal injury for someone who works on a computer four or more hours per day is nine times greater than it is for a person who spends just one hour per day on the computer.

Because these types of injuries are so prevalent in our society, Letsdoyoga.com has asked Susan Orr,

M.A., a highly experienced expert in the field of office ergonomics, to share her extensive knowledge of preventing carpal tunnel syndrome and repetitive stress injuries. In the following pages Susan will address the causes of repetitive motion injuries, offer information on how to recognize if there is a problem, and then show you how these situations can be prevented, eliminated or exposure reduced.

Howard Allan VanEs, M.A
Publisher

Repetitive Motion Injuries / Carpal Tunnel Syndrome Defined

As the name suggests, repetitive motion injuries (RMI's) are injuries that occur as a result of repeating the same motion over and over again. In the world of computer use, repetitive motions include keying, mousing, reaching, twisting or turning. Most often RMI's are caused by or exasperated by a number of interrelated factors including the period of time someone has been at a given task, lack of breaks, an individual's genetic make-up—their predisposition to handle stress, posture problems, diet, stressors in other parts of their life, and existing medical problems.

> *RMI's are caused by or exasperated by a number of interrelated factors . . .*

Repetitive motion injuries are also called by different names, including cumulative trauma disorders or muscular skeletal disorders. Symptoms of RMI's can include muscle soreness, disorders of the tendons, joints, nerves or vascular system, herniated discs or other back problems, and discomfort/pain/impairment in various parts of the body. As you can see, RMI's can be as simple as a strained muscle or as serious as irreversible physical damage.

As you might imagine an accurate diagnosis for an RMI is often challenging because symptoms can be subjective, mimic other medical problems or be interwoven with other medical issues. Below is a list of the more common "officially" diagnosed conditions of computer related repetitive motion injuries:

> *... there is actually a tunnel in the wrist, in which nerves and blood vessels run through.*

- *Carpal tunnel syndrome* is a particular type of repetitive motion injury. As the name suggests, there is actually a tunnel in the wrist, in which nerves and blood vessels run through. If there is a thickening or swelling of the tissues in or close to the carpal tunnel, there is increased pressure on the nerves and blood vessels in the tunnel resulting in the symptoms associated with carpal tunnel.

Initial complaints are usually about feeling tingling or burning in the fingers and hands, and people often talk about having dull or aching discomfort or pain. As the problem gets worse there is shooting pain, often up the arm, and less movement in the fingers, hands, elbows or shoulders. Next there is less hand strength and perhaps numbness which can result in the dropping of objects, such as coffee cup. As carpal tunnel gets more serious, there is difficultly sleeping due to the increased pain.

- *Cubital tunnel syndrome* occurs when the ulnar nerve is compressed where it passes behind the elbow in the cubital tunnel. Chronic pressure can cause inflammation and swelling of the nerve. Symptoms are weakness, numbness, tingling or pain in the hand, including the ring finger and little finger. There can also be aching in the inner forearm and elbow.

- *DeQuervains, or tenosynovitis*, is the common condition of painful inflammation of tendons constricted in the tendon sheath in the thumb and side of the wrist. The pain can migrate into the forearm. In addition to pain, there can be swelling over the thumb side of the wrist and trouble grasping and pinching. This is caused by highly repetitive hand motions such as grasping, lifting and twisting.

- *A ganglion cyst* is swelling on top of a joint or tendon sheath that usually occurs on the back of the hand and sometimes on the wrist. It is a sac of colorless fluid material that can become gelatinous over time. This fluid material, known as synovial fluid, comes from a leak in the wrist joints or the tendon sheaths. A ganglion cyst is also called a "bible" cyst. This name comes from the old fashioned treatment of whacking a bible on the hand to break the cyst. (Not recommended.) Ganglion cysts are benign, not harmful, and will not spread. However, they can cause pressure on surrounding areas and this can be painful or be considered unsightly. All lumps should be checked by a physician to make sure they are not something more serious.

- *Tendonitis* is an inflammation or irritation of a tendon, caused by overuse or injury. (Tendons are cords that connect muscle to bone.) The condition may also

f.y.i.

The point of knowing what the symptoms are is awareness—to recognize there is a problem. Then you can make the changes necessary to stop the problem from getting worse, and in the process have a great opportunity to improve your overall health, comfort and state of mind!

involve the tendon sheaths or coverings. The sheath may become thickened and restrict the movement of the tendon.

- *Tennis elbow, or epicondylitis*, is damage or tears to tendons in the area around the outer elbow from overuse or injury. It is usually caused from frequent use of the forearm muscles. There can be a burning pain on the outside of the elbow. Pain can be experienced when the palm is turned up, when flexing the wrist and fingers backward or when rotating the forearm. Difficulty can also occur when carrying a heavy load with the arm extended or gripping an object.

- *Trigger thumb/finger* is an inflamed tendon entrapped or confined in a tendon sheath. This causes the finger or thumb digit to "catch" in a bent position. It is caused by repetitive or forceful use of a thumb/finger. There is an accompanying popping or snapping sound which occurs when the inflamed tendon is pulled through the sheath when bending a thumb or finger.

Impact of Repetitive Motion Injuries

How much impact do these type of injuries have? According to the U.S. Bureau of Labor Statistics (2002), the number of occupational injuries and illnesses resulting in days away from work due to repetitive motion injuries was 1,438,194. OSHA reports that these types of injuries affect some 1,800,000 people per year and estimates that the cost to employers is upwards of $20 billion per year in workers' compensation costs. These numbers are exclusive of the additional cases of repetitive motion injuries that have not been officially

> *The Bureau of Labor Statistics reports that repetitive motion injuries for movement such as grasping tools, scanning groceries. and typing resulted in long absences from work in 2003.*

diagnosed or reported.

In Canada, a 2000/01 health survey reported that 1 out of every 10 Canadian adults had a repetitive strain injury serious enough to limit their normal activities. And not too surprisingly, repetitive motion injuries are currently being seen in a growing number of children, teenagers and college age students as they are spending more time with computers and video games.

The Bureau of Labor Statistics (www.bls.gov/opub/ted/2004/mar/wk5/art02.htm) reports that repetitive motion injuries for movement such as grasping tools, scanning groceries, and typing resulted in the longest absences from work in 2003—a median of 23 days, more than falls (14 days), transportation accidents (12 days) and fires and explosions (12 days).

What part of the body is most susceptible to repetitive motion injuries? The National Safety Council (www.nsc.org/issues/ergo/rmiwrist.htm) analyzed Bureau of Labor Statistics from 2002 and found 53% of these injuries are to the wrist and another 23% are injuries to the hand, finger and other upper extremities. Injuries to the head, neck and trunk (including the back and shoulders) accounted for 16%.

What is Ergonomics and How Can It Help?

Before offering specific recommendations it is helpful to understand the concept of ergonomics. The word ergonomics comes from two Greek words—"ergon", which means work and "nomos", which means laws. So ergonomics is literally the study of work or fitting the job to the worker.

Traditionally, a worker or user has had to adjust to the existing environment. Although time-and-motion studies were done with the invention of assembly lines, it was more from a standpoint of improving production and output than anything to do with ease of handling or comfort.

f.y.i.

For a review of the literature and historical studies in the field of industrial ergonomics, see the publication "Office Ergonomics" by Karl H. E. Kroemer and Anne D. Kroemer, published in 2001 by Taylor & Francis Ltd.

Even in the design of items used for human comfort and support (such as furniture and especially chairs) the emphasis is often on appearance or functionality. Stacking chairs are a good example of this. Although they stack great, they are miserable to sit in as any conference attendee can attest.

When human form is taken into consideration, the design is usually created to accommodate the "average" user, rather than be adaptable to a wide range of users. In valuing the effectiveness of ergonomically designed equipment and devices, keep in mind that a wider range of adjustability and flexibility will be more applicable to more people. Generally speaking, as long as there is stability and safety, a wider range is better, up to a point. That being said, too many features may lead to confusion and you may give up trying to figure out how to make it work!

In the big picture ergonomics concerns the entire environment, although there is usually an emphasis on furniture and devices to assist individuals—tools, carts or lifting devices, for example. The entire environment includes such things as lighting, noise levels, frequency of interruptions, mental stress level

> *... ergonomics isn't just what happens at work.*

and work-station placement. A thorough assessment includes evaluation of correct posturing as well as the more subjective issues such as ease and comfort.

Another point to be made is ergonomics isn't just what happens at work. The entire context of what an individual does, both at work and when not working needs to be taken into account. It is important therefore to consider and evaluate home computer workstations, hobbies and other possible contributors of discomfort or pain.

There are four major contributing factors that lead to repetitive stress injuries:

1. The problems are almost always preventable, but many people are not aware of this.

2. Although most people seem to understand the potential dangers of grosser movements (lifting a heavy item for instance) there is a lot of misunderstanding and lack of knowledge when it comes to the possible severity of injuries that result from smaller movements (mousing for example).

3. There is little understanding of the hazards associated with staying in one position for an extended period of time.

4. Symptoms signaling a problem are often dismissed or ignored. Allowing the problem to exist for an extended period of time increases severity and recovery time.

So, with a little more understanding of what can be done to prevent or reduce the severity of these injuries, a lot of pain and suffering can be avoided.

Treatment and Causes

The type of treatment individuals receive varies considerably. Sometimes people are told to just rest the affected area. There are other conservative treatments such as warm and cold soaking, physical therapy and exercise. Other types of therapy are laser light therapy (a low-energy laser light that penetrates the skin and stimulates cell activity) and ultrasound (sound waves that bombard the affected area, relaxing muscles).

Drug treatments may be prescribed such as anti-inflammatories, prescription painkillers or over-the-counter items like aspirin or ibuprofen. Cortisone (a steroid) or other anesthetic injections are used to shrink swollen tissues. Diuretics are used to reduce fluid in the body. Surgical release procedures are also used to relieve symptoms of carpal tunnel.

Prevention is the best medicine! Do not let things get to the point to where such measures are necessary. Experience has shown that some of these medical treatments are actually more harmful than beneficial! I have seen cases where the treating doctor prescribed painkillers, then released the individual back to the same scenario that caused the condition in the first place. Without making any improvements to their environment, they may cause further damage and not even recognize it because they can't feel it happening.

> *Prevention is the best medicine!*

One of the primary causes of repetitive motion injuries is, of course, repetition! If you work at a computer all-the-livelong day, you are probably engaging in a fair amount of mousing, clicking and/or keying. In certain jobs, such as computer programming, work is done primarily with the mouse. Programmers are much more apt to have repetitive motion related symptoms then those who have more variety in their computer-related work. Those who have more variety overall in their workday activities, such as filing tasks, meetings, tasks that require moving around, are even less likely to have problems.

The more work I do in the field of ergonomics, the more I believe that getting rid of bad habits is very important in solving and preventing problems. At a computer workstation, the most frequent bad habits are having a death grip on the mouse, slouching in the chair, dropping the wrists while keying and pounding too hard on the keyboard.

A person's physical condition can have an impact on susceptibility to repetitive motion injuries. It is common for those I see with computer related

Case History

I once had a client who lamented about an employee at a computer workstation who had pain, even after they had done everything they could to make her feel better. They had provided all the ergonomic equipment they knew of. When we went to see her, we took an elevator to a lower floor. The instant the elevator door opened, I knew right where she was. I could hear her banging loudly on the keyboard. My clients thought I was psychic because I walked right over to her desk without even being told where her workstation was! It wasn't hard to figure out what the problem was. ❖

problems to have illnesses such as diabetes, hypothyroidism or arthritis. Obesity is highly linked to carpal tunnel as is any condition that leads to fluid retention. Pregnant women may suffer from carpal tunnel, which goes away after the birth of their baby. Vitamin deficiency and overall poor diet are also thought to be related to carpal tunnel syndrome.

According to the National Institute of Neurological Disorders and Stroke (www.ninds.nih.gov/disorders/carpal_tunnel/detail_carpal_tunnel.htm) carpal tunnel syndrome does seem to be hereditary. People who are related are often similar in physique and hereditary conditions or predisposition to such injuries may exist.

Certainly stress plays a factor in many cases. Many of the people I see with repetitive motion related complaints have other major stressors going on in their lives, such as financial or marital difficulties. The more everyday types of stressors are contributors to this as well. There are so many deadlines and it seems

> ***f.y.i.***
>
> *Women are three times more likely to have carpal tunnel than men. (www.ninds.nih.gov/health_and_medical/pubs/carpal_tunnel.htm). This may be due to the fact that the carpal tunnel is smaller in women than men.*

everything needs to be done immediately! We work too hard and too long, drive too fast and expect too much. I heard someone proclaim, "Instant gratification takes too long". How true that seems to be. I think we also need to remember that "stressed is just desserts spelled backward".

From an ergonomics perspective mental stress translates to tension in the body, over-gripping and working in a disorganized and disconcerting fashion. Interestingly, people will adjust their posture in order to accommodate a messy workstation and in the process create more stress and risk for injury.

There are also mechanical stressors in the environment. This can come from improper adjustment or localized mechanical stressors, such as resting arms and/or hands on the relatively sharp edge of a desk. Forceful localized exertions, such as stapling, making hole punches, pulling carts or scooting chairs can cause problems as well.

Posture and positioning are very important. One thing that is often overlooked is sleeping posture. Sleeping on bent wrists/hands can cause pain and even carpal tunnel. Improper support of the body, especially the back and neck, can also lead to pain and discomfort.

Prevention

Getting the correct set-up is one of the most important things you can do, whether it be an office environment, an assembly area, or non-work related tasks such as hobbies. Let's take a look at the variables that influence the proper set-up.

Computer Workstation Set-Up and Posture

The best posture for using a computer is a neutral posture. If the chair and working surface are correctly adjusted and placed, you should be able to sit with your feet flat on the floor and your forearms level with the floor while keying. Your wrists should be in-line with your forearms and not dropped down while keying. Your arms should hang naturally at the sides, in-line with your torso without being propped up by chair arms.

There should be about a ninety-degree angle bend at your knees, hips and elbows. The chair should support your legs but not impinge on the back of the knees.

Your wrists should be in a neutral position as well, in-line with your forearms—not higher or lower than your forearms. There should be no side-to-side bending at your wrists while keying. Any posture, other than having your writs in-line and level with your forearms will impede circulation in the wrist area.

Most people don't need to use a wrist rest if they have the correct set-up and habits. In fact, most wrist rest users are wrist rest abusers! They tend to plant the wrists onto the wrist rest and then move the fingers back-and-forth from the wrist, creating a constriction in the wrist area. This isn't "resting" on the wrist rest—it is more like parking the wrists on the rest. In actuality, the wrists should float just a bit above the keyboard and the movement should come mostly from the

f.y.i.

Federal OSHA contends increasing the angle of bend in the wrist "increases the contact stress and irritation on tendons and tendon sheathes" (www.osha.gov/SLTC/etools/computerworkstatins/components_wrist_rests.htl), and they add: "This is especially true with high repetition or prolonged keying."

forearm and elbow. The caution is to avoid using the shoulders to lift the hands. The arms should be resting straight down from the shoulders and the shoulders should be relaxed. There should be no reaching forward with the upper arm.

Planting the wrists on the wrist rest is a bad habit, and one that is made worse if the wrist rest is hard or too thick. Wrist rests that are too thick sometimes cause users to sit with their shoulders lifted so they can get over the wrist rest, which in turn causes problems in the shoulder and neck areas.

> ... *most wrist rest users are wrist rest abusers* ...

Changing the habit of planting the wrists can be quite challenging. For some reason this habit seems to be especially difficult to break for those who also use a laptop or have used a laptop extensively in the past. I guess it is a bit more tempting to lay the wrists down on laptops as the keying area is sometimes more compact. If I cannot get someone to change this habit, I will put a soft, gel wrist rest, back into use. Having a soft wrist rest is better than laying your wrists all the way down onto the keying surface, which is most often a hard desk or keyboard tray.

Working Surface Height and Tilt

A workstation height should be designed so that the work is located in relation to the individual's elbow. Precise work that requires minimum strength but a clear view of the work to be done should be located approximately two to four inches above elbow height. Light assembly work requiring some strength and a normal viewing distance should be located two to four inches below elbow height. Heavy assembly type work requiring more strength and less visual clarity should be located four to eight inches below elbow height.

Placing items being worked on so that they are tilted back is a good idea for the same reason that copy holders are useful—it reduces the amount of bending at the neck to view the materials being worked with. It is easy to make what I call a production slant, which can be just a piece of wood propped up from behind. If there are concerns about the items being worked on falling to the bottom of the slant board, using even a small amount of slant can be helpful. For individuals doing a lot of reading, desktop lecterns are a good choice.

Centering

Workstations should be arranged to keep frequently used materials within an easy reach. This is especially true with heavier objects. Keep motions that are made frequently toward the center of the workstation. I can't tell you how important this is. I have seen programmers who use two monitors (for different programs running on different CPUs) develop very serious neck problems. They sit all day long and look back and forth from one computer monitor to the other. Attaching an A/B type switch, sometimes solve this. This is a switch, often

Case Study

Turning of the head seems like such a small movement, but too much of anything is well, just too much! I visited with one programmer who was facing surgery in the cervical area of her spinal column because she had worn out the discs in her neck. She related that there was talk of inserting a rod in her neck if other, less drastic, measures were not successful. As you can imagine, putting a rod in the neck would not allow for very much movement in the neck at all. ❖

foot operated, which allows the display to be switched back-and- forth from one CPU to the other.

A brief review of the spinal column is helpful in understanding why the idea of centering the tasks and movements are important. Basically you have bones that run the center of your back called vertebrae. Between each vertebrae and attached to the vertebrae above and below, are discs. The discs have some flexibility to them. The exterior of the discs are fibrous elastic bands. Inside each disc is a substance that has the consistency of toothpaste or jelly. This allows for movement in the back and neck, letting you turn and bend in all directions.

> *Keep motions that are made frequently toward the center of the workstation.*

The discs also act like hydraulic shock absorbers, so you can jump up and down without damaging the spinal column. The spinal column also protects the spinal cord, which runs down the middle of the column. Spinal nerves extend out from the spinal column along the length of the column. The spinal nerves connect the brain with various parts of the body and provide a path of communication between the brain and the body.

The discs can be damaged in various ways. Trauma, for instance, from an auto accident can cause damage to the spinal column. However, the discs can be damaged and/or worn out over time. In a sense, small tears can appear in the discs from over-loading and excessive use. You can't feel this happening and won't even know there is a problem developing. If there are enough tears in one area, the jelly-like substance in the center of the discs can bulge out into the exterior part of the disc. This results in what is sometimes called a "slipped disc", but the discs really don't slip. They are attached to the vertebrae both above and below. It is really more of a bulging or protruding disc, often called a herniated disc.

f.y.i.

It is vitally important to make sure the discs do not wear out from excessive turning and twisting resulting from not centering frequently used items in your work area or at home.

You can have a protruding disc and not know that either. But if there is some impingement of the spinal nerves, you'll know there is a problem! A disc hitting on a spinal nerve can cause a lot of pain, which is often experienced in the part of the body the nerve runs to.

Keeping things centered so that there is not a lot of over reaching or excessive stretching not only prevents disc damage, but also to allow for more natural movements. If there is a lot of reaching to the side, as there is in assembly to pick up parts for instance, there can be damage to the shoulder from excessive contraction. If leaning is required, the result can be pain in the lower back from too much stretching action in this area.

In computer workstations, I often recommend copyholders or slant boards to reduce head turning while reading hardcopy and entering data into the computer. Copyholders that are placed to the side of the computer or mounted on the side of the monitor are not very useful. These types of copyholders hold limited amounts of paper and for most people end up holding the company phone list for easy reference!

What does seem to work well are copyholders that fit between the monitor and keyboard. These are best when used with a keyboard tray because there is greater adjustability of the copyholder so that it doesn't interfere with the view of the monitor. The models that have a lip on the lower edge are best to hold papers and files in place, and ones that are wider (18 inches) allows you to move papers back-and-forth and will accommodate files). The acrylic ones let you see what items may have found their way behind the copyholder.

An additional benefit of a copyholder is the angle of the document being read. Rather than laying flat on the desktop, it is tilted up and as a result there is much less craning of your neck to read the document.

Chairs

In any workstation, it is desirable to be able to sit comfortably, with the feet resting flat on the ground. It is optimal to have approximate 90 degree angles in the body: at the elbows (so the forearms are level with the floor), at the hips (so the thighs are about parallel with the floor), and at the knees.

> *It is optimal to have approximate 90 degree angles in the body.*

There is a lot of disagreement as to the angle of the thighs to the floor. Some chair experts think it is better to be in a position where the hips are higher than the knees, but it is my opinion that this just depends on the task being performed and the person performing the task. There should not be any feeling of a need to push back in order to sit comfortably in the chair, especially at computer workstations.

If the chair seat is tilted, with the back portion of the seat pan higher than the front, it can feel like it is tipping you forward. This may feel as if you are leaning in toward your computer, and you may feel a need to push back a bit to stay seated in the chair. And when you feel your chair is tipping too far forward, you will tend to drop your wrists to steady yourself against the desktop or keyboard tray while you key, which causes a kink in the wrist and impairs natural circulation in the wrist.

In some environments, like laboratories or drafting operations, the work surfaces are designed to be used while standing and the chairs are high, to match the work surface. Because the feet then dangle above the floor, these chairs are often made with a bar in the shape of a circle—placed so that the feet can rest on the bar. It is very important that these chairs be very stable because of their extra height. For chairs with swivel casters, the more stable ones are chairs with five casters and legs that connect to a single post, mounted to the bottom of the seat pan.

Many labs and production facilities have casters on their chairs that make the chairs move too easily. When users go to sit on these chairs, the chairs can slide right out from under them. Sometimes the harder casters, often made out of hard rubber or polyurethane, are used because they can be kept cleaner (in clean rooms for example). Many times though, softer casters could be used and would make the chairs safer. Soft casters, which should be used on most harder floor surfaces, are most often made out of pliable plastic.

In offices, chairs can be made easier to move if the right casters are on the chairs. Carpet casters are designed to be used with a higher pile or a patterned rug. If the chair is difficult to move, users either digging in with the heels of their feet to make the chair move, or grab onto the edge of their desks and pull themselves in toward the desk, which can cause back problems, shoulder pain or discomfort in the hands. Also, if a chair is difficult to move, sometimes people will adjust their body, rather than move the chair. This means they may be leaning or end up in some other awkward posture that can cause various problems.

One easy fix for a chair that is difficult to move is to place a carpet protector beneath the chair. These are the plastic pieces that are used to protect the carpet

f.y.i.

In addition to slowing down the turning of the wheel on slick surfaces, soft casters have the added benefits of being more shock absorbent and are less likely to mar a floor surface.

from wear, but they can make it easier to move a chair around if they are of the right size, texture and positioned in the right place.

> **❝**
> One easy fix for a chair that is difficult to move is to place a carpet protector beneath the chair.
> **❞**

I have found most people, especially women, like to have the chair seat tilted toward the back a bit. This is often more accommodating for the hips and the back fits better into the lumbar support areas of the chair.

Men, and really anyone with a lot of weight in the upper body, do better with the seat back angled slightly backward. This is a bit of a compromise, but I have noticed that anyone broader in the shoulders or heavy in the upper body tend to fatigue forward as they tire—the shoulder girdle collapses inward and there is forward lean in the upper body. If you do lean back a bit, there is a risk of sliding down, so the backward angle cannot be too great—just enough for the weight of the shoulders to be balanced or leaning slightly toward the back.

The characteristics of a supportive chair mostly have to do with good adjustability. Many chair manufacturers now make chairs in different sizes. This is very helpful in the initial matching of a chair to an individual. It is pretty difficult to make a small person comfortable in a large chair and even more difficult to make a large person comfortable in a small chair. Once a chair is acquired that is workable, the adjustments can be made. A good chair has adjustable seat heights and has a seat pan that can be moved back and forth, as well as tilted front and back.

Chair arms are controversial. Often people performing tasks are better off with the arms removed from the chair or lowered way down. Chair arms can interfere with correct posture and hit against the work surface. Sometimes users will lean to the side to rest on the chair arms. With computer users this can be a contributing cause of carpal tunnel as leaning affects the way the keying is done and the placement of the hands. So, a good chair has adjustable as well as removable chair arms.

Look for a chair that has a back that moves up and down and tilts back and forth. It should also provide adjustable lumbar (lower back) support. For the most part, lumbar support pillows really don't work very well. Most are too thick and never seem to stay in place. The chair should be sturdy and the adjustments should hold.

Sit/Stand Stations

There are a number of situations in the workplace where the option to sit or stand while working is desired. The best way to accomplish this is with working surfaces that easily raise and lower. There are some good, hydraulic desks that operate with the flip of a switch or the push of a button available. If the work surfaces are easily adjustable, there can be two work stations, one for standing and a lower, sitting station. This requires the installation of duplicate workstation equipment-a monitor, keyboard and mouse for both the sitting and the standing levels.

I have seen workstations where everything is at the standing height and the user will add a very high chair. This does bring up the issue of the stability of the chair, which is of great concern with individuals who are already experiencing back problems, as well as concern about the safety of getting into and out of a higher chair.

f.y.i.

People with back pain sometimes feel better if they can stand part of the time.

Footstools

In an office environment footstools are seldom needed. Usually I see footstools kicked way back under the desk, out of the way. They are a hassle to use and only work if the chair is directly in front of the footstool. If there is an adjustable surface height or keyboard tray and a chair that is height adjustable, most of the time it is possible to get the set-up arranged so that you can sit with your feet flat on the floor.

Footstools are useful if there is no way to get your feet flat on the floor. This sometimes happens when your chair lacks adjustability and the user is small in stature. Footstools are a good idea if you have knee or back pain and want to have the option of stretching out your legs.

They are also a good idea in situations where a lot of standing is required. Placing one foot up on an object while standing provides some flattening of the lower back area and for many people, this provides a change in posture and some experience relief in the lower back. Be careful though, as stools can be a tripping hazard. Some of the work areas I have visited, have rails, as is seen in bars, to rest your foot on instead of a footstool.

Keyboards

f.y.i.

The seminal work on basic information for choosing a keyboard type is a PDF at the NIOSH (National Institute of Occupational Safety and Health) website- www.cdc.gov/niosh/97-148.html.

In my seminars on office ergonomics, the subject of alternate keyboards often comes up. There haven't been many independent studies on the various types of keyboards and this subject gets everyone very confused. There are some very exotic forms of keyboards available and this makes it even harder to understand. I recommend three or four forms of keyboards which meet most needs.

For most users the basic keyboard that comes with computers is fine, so long as it has a good, springy response in the keys. That being said, the design does have an interesting flaw as there are all these keys on the right-hand side that have to be cleared when reaching for the mouse. When you consider most users are right-handed, it makes you wonder what the designers were thinking. It may have been that initially, no mice were used and so everything was put on the right because of the fact most users are right-handed.

For those who are smaller in stature, or those who have extensive movement back-and-forth from mouse to keyboard, I sometimes recommend what is called a mini or compact keyboard. This works for those who do not utilize the right-hand 10-key. The other keys on the right side (for example the "page up" "page down" keys or the "home" or "end" keys) are usually integrated into the keyboard so that there is a much smaller footprint. Not only does this allow for a smaller keyboard, but also brings the mouse much closer to the keyboard and helps to center the work activities as discussed earlier.

> *Keyboard trays are useful in getting the right keying height.*

Additionally, keyboards without the right side 10-key can be angled up at various heights in the center and allow for good adjustability. Many of the keyboards without the 10-keys also have separate 10-key pads that can be purchased and used when needed. When not in use, the 10-key can be moved out of the way so the mouse can be brought closer in, resulting in less opening up of the mousing shoulder.

For users who have to bend their wrist inward to get their hands on the keyboard (usually users who are wider at the shoulders) the so-called "natural" keyboards are good, so long as their shape is of the smaller style. Natural keyboards are usually slightly mounded at the center and often split open a bit at the front center of the keyboard. There are a lot of offices that acquired these "natural" keyboards for all their employees, when in fact they are not good for most women and anyone with a small width in the shoulders.

Over the years, many individuals have become used to this style keyboard in the belief that they are more ergonomically correct. There is a pretty big reach to the outer keys. This is especially true for those models with additional keys and features, such as the internet versions with various programmable buttons.

Keyboard trays are useful in getting the right keying height. They are also useful in helping you back-up from the monitor when there isn't enough space on the desktop for the monitor footprint. They can also be useful with taller individuals when the tray raises higher than the existing work surface.

There are all kinds of keyboard trays. An important characteristic of a good keyboard tray is that it is very sturdy, which usually means it is screwed into the underside of the desk. It should also have good adjustability and the adjustments should hold. Good adjustability means that it is easily adjusted up and down, in and out, and tilt forward and back. Although you can't get too much of a tilt before you get a runaway mouse, sometimes a little bit helps a lot.

Adjusting a tray to have a bit of a negative tilt (lower in the back than in the front) allows for additional knee space, and your hands fall more naturally onto the keys. Negative tilt doesn't work well with the "ergonomic" or mounded keyboards, as they already have a larger, mounded, wrist rest area, and this just raises it higher.

Another feature of a good keyboard tray is that it does not have any hardware hanging down underneath it, which can bang your knees on. Any mouse tray should be at the same level as the keyboard tray and have no lips or other impediments on the tray that will cause you to have to constantly lift and lower your hand. You want the movement to be easy and level.

The worst kinds of trays are the plastic trays that attached to the desktop with clips and don't adjust, the type with a lip between the keyboard tray and mouse tray, the kind that have the mouse tray attached at a different level, or any tray with limited adjustability.

Mice

The mice that are now available are much better than in the past. A lot of commercial users still have the smaller mouse that comes with the computer. These types of mice are not really made for heavy use. They are more for those who do occasional mousing. They just don't fit the hand well and don't provide any support or ease of movement. It is critical to have a good mouse for anyone who does more than an hour or so of mousing every workday. Look for a mouse that moves easily and fits your hand well. Fitting your hand means that it provides support so that your hand isn't arched up in the air to grip the mouse.

A good mouse provides structural support for your hand so that it is supported by the structure of the mouse and can rest easily on it. Also, a mouse or track ball shouldn't be driven from the base of the finger, but instead should be operated with the fingers or the palm of the hand, depending on the type of track ball.

Optical mice move very easily and have fewer maintenance issues. Makers of the newer laser mice (www.logitech.com) claim the laser mice have "20 x

> *Look for a mouse that moves easily and fits your hand well.*

more sensitivity to surface detail-or tracking power" than optical mice. I would avoid thumb-operated track balls because they tend to lead to thumb/hand problems with overuse.

Wireless mice are good for eliminating problems with tail drag and many now have chargers the mouse sets in when not in use. (This eliminates the previous problem of running through a lot of batteries.)

Programmable mice are good for those who have a lot of repetitive movements in their keying and those who do a lot of cutting and pasting. Prior to programming the mouse, you will need to understand which actions would benefit most from being programmed. Having good knowledge of exactly what you are doing during the time you are working on the computer will help you accomplish this.

Of the programming I have seen done on the mouse, the most frequently programmed function has to do with cutting and pasting. For those who do a lot of this activity, it can be very helpful to program the mouse for this function. Without programming, the user is required to click on the object to be dragged. Then, while holding the clicker button down, the user must drag the object to the desired spot before the clicker is released and the object is dropped.

With programming, you just click on the object to be dragged and then move it to the desired spot, click again and the object is released. There is no need to keep the clicker button down during the dragging of the object.

If your work requires extensive mousing and you would rather use key strokes for some of the functions normally performed with a mouse, keystroke shortcuts may be helpful. An example of this is, instead of highlighting text, then clicking on "Edit" and "Copy," you can highlight, and then hit "Control" plus the letter "C."

Note that while these keystroke shortcuts may reduce mousing time, they do increase keying time, so if you are doing more keying than mousing, these may not be desirable alternatives to mousing.

You can also use macros for tasks that you do frequently. A macro is just a set of computer instructions you can connect to shortcut keys or to a name you give to it (the macro). Microsoft provides instructions for Office XP (http://office.microsoft.com/en-us/assistance/HA010192301033.aspx) and states that you save time

f.y.i.

Here is Microsoft's sheet "Keyboard Shortcuts for Windows" website: http://support.microsoft.com/default/aspx?scid=kb;en-us;q126449

when you replace an "often-used, sometimes lengthy series of actions with a shorter action".

I don't often recommend the more exotic types of mice, such as the trigger mice, which are gripped like a joy stick, with the hand on the edge of the outside of the palm and the clicker pushed with the thumb. The trigger mice seem to encourage bad posture as users often have their arm stretched out straight and slide down in their chairs.

Other "natural" posture mice are similar to the trigger mice in that the hand is more in the handshake position, on the side. The problem with these mice is there is a much larger movement required to go from keyboard to the mouse and back. Not only do you have to get up and over the mouse, but there is also a turning of your hand.

Many people report positively on the writing tablet or pen tablet mice. These are basically a flat writing surface with a stylus. They seem to work especially well for those who do a lot of design or graphics work. Often, however, they create the same problem experienced by people who do a lot of writing—extensive gripping and pinching of the stylus with heavy usage. Having a wider and softer grip on the stylus helps some users.

Computer users who have the most problems with mousing are those who use the mouse extensively, especially those who work long hours and don't take frequent breaks or add variety to their daily activities. And the problems can be pretty devastating. Not

Case History

A few years ago I took a night class and became friends with a lady in the class who had worked for many years for a large computer manufacturer. For many years, she had spent the better part of her workday on the computer. During the years that her carpal tunnel developed, there wasn't much information available on this condition and how to prevent it. She just kept on working, even though the problems were progressing. Having retired on disability retirement, she hadn't worked in years when I met her. This woman could barely grip a pen to take notes in class. She related that she had a lot of pain, even though she hadn't used a computer in a long time. Don't let this happen to you. Use this story as motivation to take the steps necessary to protect your wrist from carpal tunnel syndrome! ❖

everyone recovers completely from repetitive motion injuries and carpal tunnel syndrome from extensive mousing seems to be especially debilitating.

Monitors

f.y.i.

The ideal monitor height varies according to whether or not you wear corrective lenses. If you have good vision, your eyes should be no lower than the top one-fourth of the viewable screen when you are looking straight ahead with your chin level with the floor. This can be altered somewhat if there is a program or activity that requires looking at a particular portion of the screen most of the time. Some programs actually have you spending most of your time looking at data on the bottom half of the screen and in this case the screen should be slightly higher.

More and more experts seem to be lowering the recommended height for monitors, with many advising that your eyes be level with the top of the monitor.

People wearing progressive lenses may actually be looking at the computer monitor through the bottom half of their glasses. If you are one of these people, place the monitor much lower so you don't have to sit with your head tilted back all the time.

This is a bit of a common sense item but it is worth mentioning if there is any awkward position or craning of your neck, something needs to be adjusted. It is good to have a slight downward tilt to your chin. It is harder on the neck to have to be straight up all day than it is to have a slight downward angling of your head and chin.

The proper distance from the monitor is somewhat of an individual measurement. It depends on your eyesight, the task at hand, font size and general ease of readability. Avoid leaning forward or craning your neck to see the monitor. By the same token do not be too close in. The general rule use to be an arm's distance, but there really is no hard and fast rule for this item. I often deal with folks who are working on code and will have very large screens and often more than one monitor. So, in addition to the previously mentioned problem of extensive turning of the neck, they also have a larger height and width of monitors, making appropriate placement and height more difficult to discern.

Vision/Eye Strain

> ❝
> ...users do not notice when monitors have gotten older and need to be replaced.
> ❞

The clarity of monitors is improving. One thing I have noticed is that users do not notice when monitors have gotten older and need to be replaced. Loss of clarity and brightness happens so slowly, it just isn't noticed. The monitors tend to get darker and fuzzier with age.

There are a few things that can be done to relieve eyestrain from computer use:

- Check to see if the monitor needs to be repositioned to reduce glare. The glare screens that are now available also work well. Adjust the contrast and brightness settings on your monitor to compensate for the existing lighting in the room. (This can be done from the "Control Panel", then "Display" and may be adjustable directly on the monitor controls.)

- Blink your eyes often and take a break by alternating tasks when necessary.

- Occasionally focus on objects 20 feet or more away.

- Check with your eye care professional to see if special lenses are needed for computer work.

- You may need to consider adjusting the font size. (From a Word document, go to "Format" then "Font" to adjust font size. On the Internet, go to "View" then "Text Size.")

- Close your eyes and breathe deeply for 30-60 seconds.

- Place material as close as possible to the monitor to avoid frequent head movements.

- Adjust the monitor settings for comfort/ease of viewing.

- We don't blink as often when looking at a computer monitor as when reading hardcopy. Also, most office environments have very low humidity, so it is common to end up with dry, tired eyes. Eye moisturizers are sometimes recommended. However, there are precautions (http://my.webmd.com/content/article/64/72258.htm) on the use of vasoconstrictors. The concern is once the vasoconstrictor drugs wear off and the blood vessels relax, they may dilate, causing more redness. This is a phenomenon known as "rebound redness." Preservative-free artificial tear supplements are recommended instead of vasoconstrictors.

- To maintain healthy vision, adjust the refresh or redraw rate on your monitor, if you have a CRT (cathode ray tube) monitor, not an LCD (liquid crystal display) monitor. This rate is the number of times per second the monitor screen is refreshed. Slower refresh rates can cause flicker at higher rates of resolution. Flicker stimulates the eyes to refocus, tiring the eyes with repeated refocusing. On most computers, this change can be made by accessing "Display" through the control panel, then "Settings" then "Advanced" and finally "Monitor." Check your monitor instruction manual to check the frequency and resolution your monitor can support before making any changes.

- Lighting levels in the general work area can be a little lower if there is computer use only, but may need to be a bit brighter or have task lights added if there is some hardcopy document viewing. The American Industrial Hygiene Association indicates that some

f.y.i.

Microsoft Support, on their website, (http://support.microsoft.com/kb/q228212) reports that although some people perceive screen flicker where others do not, most people perceive no screen flicker if the refresh rate is 72 hertz or higher.

> *To maintain healthy vision, adjust the refresh or redraw rate on your monitor.*

experts say that the background lighting or source document (the hardcopy the user is reading while keying) be no more than three times brighter than the monitor (www.aiha.org/GovernmentAffairs-PR/html/OOergo.htm).

- You may also want to experiment with text and background monitor colors. While I have found most people report a black text on a white background is the most comfortable, there is a high degree of variability of color perception from individual to individual. The American Industrial Hygiene Association says a light screen background (dark type or images on white or pale background) is easier on the eyes. (www.aiha.org/GovernmentAffairs-PR/html/OOergo.htm).

Awkward Postures

The most troublesome awkward posture from an office ergonomics perspective involves cradling the phone between the shoulder and the neck. This causes extensive side bending in the neck and hunching of the shoulder. If you are using your computer (keying) while doing this, it also changes the keying posture of the hands. Even if there isn't much phone use, there should be consideration for the use of a headset. And of course defiantly used if there are extended phone conversations. The headsets available now are lightweight and easy to use. You'll just need to make sure you have one that is comfortable, otherwise the rate of usage decreases.

Extending the neck to see the monitor through bifocals or progressive lenses has already been

Case History

I was working with a publishing company years ago that did not have any headsets available. Many jobs there required a great deal of keying and as a result they had a lot of employees with repetitive motion and carpal tunnel type symptoms. When we placed headsets in use, the number of employee complaints of stiffness, numbness or pain in their hands, arms and wrists decreased by almost half! This is amazing to most who hear this story. They can't understand how not using headsets can contribute to neck and shoulder problems, but the connection to the change in the posturing of the hands and arms is critical. ❖

mentioned, as has craning of the neck to see hardcopy target documents on the desktop. Other examples of awkward postures include overreaching for files and documents, twisting and turning of the torso while reaching objects or while just sitting at the computer. Obviously, these moves should be minimized to avoid injury.

Working in the corner of a desk creates some difficulties and many people just can't seem to get squared to their workstation in desk corners. If a keyboard tray is used, you can end up trying to mouse underneath the desktop and twisting and turning the mousing arm to reach the mouse. One of the best solutions for this is to add a corner diagonal plate. These are usually made of metal and fit like a sleeve over and under the desktop. They are then attached to the desktop and provide for a squaring off of the corner. Many corner diagonal plates have pre-drilled holes for attachment of a keyboard tray.

Laptops

From an ergonomic standpoint, laptops are not designed well. The keyboards are too small for most users, they sit too low on the desk and require you to scrunch down to key and see the monitor. The various types of mousing devices are seldom very ergonomic in design. Because laptops are mobile, they are often used in situations where it is difficult to get the appropriate adjustments for comfortable computing, such as on airplanes or in hotel rooms. The less often these compromised positions are utilized, the better. As a general rule, laptops should be used sparingly.

> *As a general rule, laptops should be used sparingly.*

The more attachment accessories you are able to hook into the laptop, the better. I usually recommend a separate, more ergonomically designed mouse, a full-sized keyboard and a separate monitor. Docking stations are very helpful. Some users like the clarity of laptop screens, and that's okay, so long as the monitor is at a height that allows for easy viewing of the monitor.

Fixes that Often Don't Work

I cringe when I hear people tell me they were advised by ergonomic or health care professionals to switch the mousing hand. This happens when they are having problems with their dominant mousing hand. The thinking is you can give the dominant mousing hand a rest by using the other hand. The concern with this is that the conditions that caused the problem in the dominant mousing hand may have not been solved. The problem is just transferred to the other hand and you may end up with bi-lateral carpal tunnel! And there is usually a quick transfer of the symptoms from one side to the other. You need to be really careful with this and make sure the set-up has

> *I cringe when I hear people tell me they were advised by ergonomic or health care professionals to switch the mousing hand.*

been adjusted to provide the most ergonomic posture and equipment.

Even with a thorough analysis of the set-up and a review of the factors contributing to the condition in the dominant mousing hand, I still have a reluctance to recommend switching mousing hands as a remedy. There may be unrecognized factors involved and there is little understanding of the mechanisms that facilitate the development of symptoms in the non-mousing hand or in the quick transfer of symptoms and impairment when there is a switch in the mousing hand. This option is a better choice for preventing problems, rather then a possible cure for existing discomfort.

Many people have tried voice recognition software. It keeps improving, but the word I get is it just isn't there yet. The time it takes to adjust the software to the individual and the individual to the software is pretty extensive. Also, many tell me that they spend more time fixing what was typed than it would have taken to just type it!

Foot operated mice are sometimes used by people with extensive carpal tunnel. The problem is, guess what, they end up with repetitive motion injuries in the feet and ankles!

I am not a big fan of the use of arm braces while working at the computer. Those who use them say they think it helps to keep their wrists and forearms level with the floor. However, the braces can cause compromised posturing when the employees put their fingers around the brace itself. Although the braces have improved, for the most part they just don't work well. Braces do work in one instance—for those who have a habit of sleeping on bent wrists and hands which can restrict the flow of fluids and nutrients in the body. Because you can't do much about positioning habits while asleep, braces may prevent bending of the hands and arms while sleeping.

Behavioral Prevention Tips

So what else can you do to prevent repetitive stress injuries and discomfort?

- Certainly avoid or reduce repetitive motions as much as possible. Keep the wrists and forearms straight while keying.

- Take micro breaks and stretch and exercise during these breaks. The National Institute for Occupational Safety and Health (www.cdc.gov/niosh/restbrks.html) reported, "Adding short breaks through the day may relieve cumulative discomforts from repetitive motions and static postures in a way that conventional break schedules do not." The study they did compared data entry workers who had just two fifteen-minute breaks a day, to those who supplemented this break schedule with four five-minute breaks. The result was, those with the added short breaks ". . . consistently reported less eye soreness, visual blurring, and upper-body discomfort under the supplementary schedule."

- Taking breaks or occasionally changing positions also has the benefit of improving the health of the spinal column. There is an interesting book by Galen Cranz, entitled simply "The Chair." She says changing positions is "essential for the health of our spinal discs" because ". . . they don't have veins, so must get their nutrition via a process of diffusion, which depends on a pump or sponge mechanism. This requires alternately overloading and underloading the spine through movement."

> *Take micro breaks and stretch and exercise during these breaks.*

- Avoid sleeping on your hands and improve diet and conditioning with exercise.

- Also, begin counter measures in early phases before damage progresses.

- Engage in "comfortable computing." This includes sitting upright and facing the computer straight on, with the hands and wrists straight while keying.

- Adjust the chair height and seat back and sit with the feet flat on the floor.

- Touch the keys lightly and consider programming the mouse. Use keyboard short cuts when appropriate and utilize macros for often used or lengthy keyboard combinations.

- Keep the equipment well maintained. Get the monitor at the appropriate height and use document holders and slant board, when needed.

- Get a headset if there is any amount of time spent on the phone.

- Take a look at how things are set up at home. You can have a great set-up at work, but if you go home and work on a computer at home that isn't properly set-up, this can contribute to problems.

- Be careful when engaging in activities other than computer work. People who work in sedentary jobs are more susceptible to injuries from activities such as lifting, because they just spend too much time sitting and aren't in the greatest physical condition. Become aware of the proper techniques for lifting, such as, thinking through the lift before you begin, bending the knees and using the legs to lift, grasping the object firmly and keeping it close to you and lifting with a smooth, controlled motion. Turn your feet, instead of twisting your back and know you limits. Also try to limit the weight and size of the objects being lifted, as much as possible. Use mechanical lifting equipment or just get help with large, heavy or awkward objects.

Websites

Ergonomics—www.lni.wa.gov/IPUB/417-133-000.pdf

This is probably my favorite document on office ergonomics. It was published by the State of Washington for reference by Washington State employers and employees. It is the most comprehensive and best document on the subject of office ergonomics I have seen. There is very little I find to disagree with in this document.

Ergonomics—www.cdc.gov/niosh/epintro.html

This is a great site on office ergonomics by the National Institute of Occupational Safety and Health.

Lab Ergonomics—www.cdc.gov/od/ohs/Ergonomics/labergo.htm

Where you want to go to know more about ergonomics in laboratories and is a Center for Disease Control site.

www.handupperex.com/common_problems.htm

This is a great site for information on common problems in the hands and upper extremities.

www.ninds.nih.gov/disorders/carpal_tunnel/detail_carpal_tunnel.htm

The National Institute of Neurological Disorders and Stroke (part of the National Institute of Health) has good information at a site entitled "Carpal Tunnel Syndrome Fact Sheet."

www.tifaq.org

The Typing Injury FAQ site has great information.

www.osha.gov/SLTC/etools/computerworkstations/index.html

A good Federal OSHA site (U.S. Department of Labor, Occupational Safety and Health Administration) that has user-friendly information.

OSHA Regulations

www.osha-slc.gov/SLTC/ergonomics/index.html

The Federal Ergonomics Standard was overturned and it is unknown if and when such a standard might be implemented. Fed OSHA does however provide guidelines. They state they will enforce ergonomic violations under the general duty clause, which requires employers to provide a workplace free from hazards.

www.osha.gov/fso/osp/

There are some states with their own OSHA regulations and you can locate these at this Federal OSHA site.

www.dir.ca.gov/Title8/5110.html

California has had a concise ergonomics regulation on the books for several years.

www.dir.ca.gov/DOSHPol/P&PC-13.HTM

The enforcement of the California ergonomics standard is outlined in the Cal-OSHA "Safety and Health Policy and Procedures Manual," utilized by Cal-OSHA field enforcement personnel.

Associations & Publications

www.iea.cc

The "International Ergonomic Association" is a worldwide association of ergonomics and human factors society.

www.hfes.org

The "Human Factors and Ergonomics Society" is a US organization. They publish periodicals entitled "Human Factors" and "Ergonomics in Design."

ww.aiha.org/SplashPages/html/topic-ergonomics.htm

The American Industrial Hygiene Association has an "Ergonomics" page. Also, on this page they have one of the few Spanish publications on-line I have seen that provides an overview of ergonomics, entitled "An Ergonomic Approach to Preventing Workplace Injury."

Ergonomic Computer Equipment and Accessories

Economical Chairs

Haworth: *www.haworth.com*

OfficeMaster: *www.officemaster.com*

Global Chairs: *www.globaltotaloffice.com/usa2004*

Alternate Chairs

Herman Miller: *www.hermanmiller.com*

Steelcase: *www.steelcase.com*

Allsteel: *www.allsteeloffice.com*

Chairs for Tall Users:

Steelcase: *www.steelcase.com*

Neutral Posture: *www.neutralposture.com*

Body Built: *www.bodybilt.com*

Chairs for Users Over 250 Pounds:

Steelcase Criterion Plus: *www.steelcase.com*

Neutral Posture: www.neutralposture.com

Body Built: www.bodybilt.com

Adjustable Height Desktops:

Alimed: www.alimed.com

I Go Ergo: www.igoergo.com

Source Ergonomics, Inc.: www.source-ergo.com

BOSTONtec: www.bostontec.com

Keyboards

Alimed: www.alimed.com

Buy.com: www.buy.com

Comp USA: www.compusa.com

Kensington: www.kensington.com

Microsoft: www.microsoft.com

Keyboard Trays

Alimed: www.alimed.com

Humanscale: www.humanscale.com

KIIKA: www.kiika.com

Workrite: www.wrea.com

Mice and Alternate Input Devices

Alimed: www.alimed.com

Kensington: www.kensington.com

Logitech: www.logitech.com

RollerMouse: *www.contourdesign.com*

RollerMouse: *www.pcmall.com*

3M: *www.3m.com*

Wacom Pen Tablets: *www.wacom.com*

Footrests

Alimed: *www.alimed.com*

Workrite: *www.wrea.com*

Humanscale: *www.humanscale.com*

Monitor Arms

Alimed: *www.alimed.com*

Workrite: *www.wrea.com*

Humanscale: *www.humanscale.com*

Wrist rests

Alimed: *www.alimed.com*

PC Mall: *www.pcmall.com*

3M: *www.3m.com*

Headsets

Hello Direct: *www.hellodirect.com*

Plantronics, Inc.: *www.plantronics.com*

Headsets.com: *www.headsets.com*

Document holders

Alimed: *www.alimed.com*

Airtech: *www.airtech.net*

Humanscale: *www.humanscale.com*

Vu-Ryte: *www.vu-ryte.com*

Workrite: *www.wrea.com*

Voice Recognition Software

Dragon Naturally Speaking: *www.dragontalk.com*

Bonus Section: Exercises to Relieve Stress at Computer and Workstations

By Howard Allan VanEs, M.A.

Before beginning any program of exercise it is a good idea to check with your physician, especially if you have any physical challenges or injures. *Remember—if it hurts don't do it. The goal is a mild stretching sensation not pain!*

Belly Breathing—good for reducing stress and improving posture

Your breath, posture and energy are intimately linked. Think about it for a moment. When your breathing is shallow your energy is poor. Your chest most likely will be collapsed, shoulders forward and lots of tension in the shoulder and neck area. Conversely, when you are relaxed, your energy is better and breathing is deeper—from your belly (diaphragm). Your posture will also be more upright with your shoulders back and down and your chest open. Your energy, posture and how you feel in your body can be easily influenced by becoming conscious of your breathing patterns and changing them accordingly.

Below is a breath that will improve your posture, energy and release tension in your torso. It is advisable to take a simple breathing break every hour or so while at work to get the most benefit from this exercise. The challenge is to remind yourself to do it!

Place your hands on your belly. As you inhale let your belly expand as you inhale taking the breath first into your abdominal area and then up into your chest and shoulders. As you exhale release the breath from your chest then your belly—pressing your belly in with your fingers. Repeat a few times with your hands on your belly. Once you think you have the rhythm of this you can do it without your hands. If is difficult for you to do, just focus on your belly moving out on the inhalation and in on the exhalation. Take 5-10 long slow breaths. You will find this breath to be even more relaxing if you can slow the exhalation, making it longer than the inhalation.

Eyes:

Note: Focus on long slow breaths with all exercises

1. Keeping your head level and straight: inhale as you look to the right and then exhale as you look to the left—repeat 5 times
2. Keeping your head level, inhale as you look up and exhale as you look down—repeat 5 times.
3. Inhale as look down to the left and then exhale as you look up to the right—repeat 5 times.
4. Inhale as you look down to the right and exhale as you look the up to the left—repeat 5 times.
5. Slowly rotate your eyes clockwise 5 times and then counter clockwise 5 times.
6. Rub your hands together until they are warm and place your palms over eyes. Leave them there for a few seconds and then as you exhale slowly slide fingers down over cheeks. Repeat for a total of 3 times. This is very soothing for your eyes and very relaxing overall!

Neck:

1. *Head turns:* Inhale and turn your head to the right. As you exhale turn your head back through center to the left. Inhale here and then as you exhale turn you head

back through center to right. This is one round. Go slowly. Repeat for a total of three to five rounds.

2. *Head up and down:* Inhale as you lift your head up gently and as you exhale release your head towards your chest. Repeat for a total of three to five rounds. (If you have a neck injury or have pain in your neck do not do this exercise.) On the last round keep your head at your chest. Then as you inhale bring your chin towards your right shoulder and nod a few times. Release and then do the left side.

3. *Side of neck:* Extend you left arm straight out to the side at shoulder height. Place your right hand on left ear and gently pull to the right. Take five breaths and repeat on the other side.

Shoulders:

1. *Shoulders shrugs:* Inhale, bring your shoulders up towards your ears, hold for a second or two. As you exhale release your shoulders with a loud (vocal) exhale. Repeat five times.

2 *Shoulder rolls:* Roll your shoulders forward five times and then back five times. Coordinate the movement with your breath—go slowly.

3. *Back of shoulder stretch:* Bring your right hand across your left shoulder bending your right elbow. Bring your left hand to your right elbow and push your elbow. Take five breaths and do the other side.
4. *Overhead shoulder/arm stretch:* Sitting upright, interlace your fingers with palms facing away and take your arms overhead pressing your hands towards the ceiling. Lift strongly through arms. Take 5 breaths.

Wrists, Hands, and Fingers:

1. *Hands and fingers stretch:* Extend your arms in front of you to shoulder height. Inhale and spread your fingers wide. As you exhale make a fist. Repeat five times.
2. *Wrist release:* Keeping your arms in front of you at shoulder height and hands in a fist—slowly roll your fists to the outsides making circles in the air with your hands for five rotations—then go back the other way for five rotations.
3. *Fingers, palm, and forearm stretch:* Extend your left arm straight out in front of you at shoulder height with your palm facing up. Place your right hand on fingers of left and pull your fingers back towards your torso. Take five breaths and do the other side. This can be an intense exercise—if there is any pain, ease up or don't do it.

4. *Fingers stretch:* Grab your right thumb with your left hand (gently pull) on your thumb as you slowly rotate it in one direction a few times then back the other way. Repeat with each finger on both hands.

Back and Chest

1. *Chest and back stretch:* Sitting at the edge of your chair seat take your arms to the sides of your body with palms facing forward. As you inhale take arms back and as you inhale take them forward—bringing your hands together. Go slowly. Raise your arms a little higher and repeat the same movement. Repeat 3 more times raising your arms a little more each time until they are slightly higher than shoulders. Then begin to lower your arms in the same way repeating the process over 5 breaths—bringing your hands by your hips to rest. This exercise also improves breathing and will help build your energy.
2. *Chest and front of shoulders stretch:* Stand up with your feet hip width apart. Lift up through your spine and the sides of chest. Bring your arms behind back and interlace fingers. Draw your arms straight and lift them slightly keeping your torso upright. Don't let your chest collapse. This exercise is great for your

posture, especially if your chest tends to collapse and your shoulders move forward towards each other.

3. *Seated twist:* Sit at the edge of your chair seat and lift up through your spine. Cross your right leg over left. Sit forward slightly keeping your spine long. Bring your left arm across your right thigh and take your right arm behind you onto top of chair back. Twist to the right. Hold for 5 breaths and do the other side.

Legs and Hips:

1. *Knee to chest:* Sit at the edge of your chair seat. Bring your right knee in towards your chest. Place your hands on the back of your hamstrings (back of thigh) and pull towards your chest. Hold for five breaths and switch to the other side.

2. *Hip stretch:* Sit at the edge of your chair seat with your legs hip width apart. Place your right ankle on your left thigh close to knee. Lift up through your spine. From your hip crease bring your torso forward. Do not force your torso forward—just go as far as gravity takes you. If your hips are too tight to do this, bring your right mid-shin area onto your left thigh instead. Work with your breath and your body may relax more. Take five breaths and do the other side.

3. *Hip and lower back stretch:* Sitting at edge chair with legs hip width apart. Fold your torso fold forward between legs moving your upper body closer to the floor. Take your hands to floor and experiment with walking hands forward and back between your legs. Hold for five deep breath and come up. Caution—if you have any lower back challenges or feel any discomfort in you lower back do not do this exercise.

Rest:

From a seated position place your forearms on your desk and then place your forehead on your forearms. Slide your chair back so that your torso is at a bit of an angle to the floor—you want to feel a slight stretch in the sides of your body. Close your eyes and breathe slowly for 3 - 5 minutes. In addition to being very relaxing this is a great stretch for your upper body.

Also Available
from Letsdoyoga.com

Order any of the items on the following pages online at www.letsdoyoga.com *or use the order form at the end of the book.*

Tight Shoulder Relief

Howard Allan VanEs, M.A.

96 pages $14.95

Do you suffer from tight shoulders or neck area? If so, you are in good company! According to the National Center for Health Studies 13.7 million people visited a doctor in 2003 for a shoulder problem. This number doesn't include those that have shoulder problems but haven't seen a doctor. In fact, many health experts suggest that most of us will have some sort of shoulder issue by the time they reach 65.

Tight Shoulder Relief explains why your shoulders get tight, shows you how to relieve them, and prevent future problems. Easy to follow exercises help you reduce discomfort, improve flexibility and have you feeling good in your body again. The anatomy of the shoulder is also provided for reference.

This book is a must for those who are at a high risk for shoulder problems: computer users, dental hygienists, gardeners, hair stylists, and professional athletes, etc. If your job requires a lot of repetitive motions your will find Tight Shoulder Relief to be an important resource for reducing pain and keeping your body healthy.

Features

- Over 50 different exercises to relieve tight shoulders with photographs and easy to follow instructions
- How your shoulders move and why they get so tight
- Shoulder Anatomy 101
- How breathing effects your posture and shoulders

Beginning Yoga: A Practice Manual

Howard Allan VanEs, M.A.

192 pages, $24.95

An essential resource for beginning through intermediate yoga students! **Beginning Yoga: A Practice Manual** has been carefully designed to help students develop a solid foundation in yoga through their home practice. This is a clearly written, easy to use book with just the right amount of information to support your practice. And it's large 8 1/2 X 11 format with full-page photos makes postures easy to understand and reference. A thoughtful balance of theory and practice are presented to help you ground into the philosophical aspects of yoga while you deepen and refine your physical practice.

Features:

- 50 postures with full-page photos and clear step-by-step instructions
- How to practice at home—over 20 follow-along practice sessions
- Specific practices for energy, relaxation, and preventative back care
- Introduction to Meditation
- Introduction to Pranayama (yogic breathing)
- An overview of yoga's history and philosophy
- Lay-flat bookbinding stays open for easy reference

Shavasna / Deep Relaxation (Audio CD)

Howard Allan VanEs &
Debra Marcus
Length: 74 minutes
Price: $16.95

There is a profound place of peace within all of us and the ancient practice of yogic sleep opens the door to this place of stillness and well-being. Shavasana is an inner retreat—a delightful break from the demands of our stress filled lives. As you listen to this CD and experience the benefits of deep relaxation you will feel renewed and refreshed. A great stress buster! Four different shavasana experiences are on the CD:

Body Scan and Shavasana: The power of imagery and breath is combined with a classical shavasana experience to help send the message of relaxation to every area of your body. As your tensions dissolve, you'll discover the deep peace that is at your core.

Progressive Relaxation and Body Scan: The proven stress reduction practice of progressive relaxation is combined with a body scan and shavasana for an extraordinarily deep experience of peace and relaxation. Become aware of where you hold tensions so that you can consciously choose to let go of them and experience the bliss underlying it all.

Beach Shavasana: Imagery of a warm sunny day on a beautiful white sand beach combine with the kinesthetic sensations of gentle undulating tides rolling over your body massaging away tensions—leaving you peaceful and at ease. Break the cycle of stress and enjoy a mini-vacation for your mind and body.

Classical Shavasana: A brief deep relaxation experience that will quiet your mind and body leading you directly to stillness and a sense of peace. Come back to your day feeling renewed and revitalized.

Office Ergonomics: Preventing Repetitive Stress Injuries and Carpal Tunnel Syndrome

By: Susan Orr

Edited by: Howard Allan VanEs

104 pages $14.95

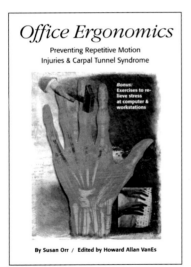

Did you know that 90% of the workforce in the U.S. uses a computer? And did you know that the more you use a computer, the more likely it is that you will have a repetitive stress injury? In fact, the risk of a muscle-skeletal injury for someone who works on a computer four or more hours per day is nine times greater than it is for a person who spends just one hour per day on the computer. *Office Ergonomics* identifies the causes of repetitive motion injuries, provides you with information on how to recognize if there is a problem, and then shows you how these situations can be prevented, eliminated or exposure to them reduced.

Features:

- 100s of tips for setting up your workstation to make it comfortable, efficient and reduce the risk of injury
- Treatment and causes of repetitive stress injuries
- Positions and setups you must avoid
- Fixes that don't work
- Behavioral prevention tips
- How ergonomics helps
- Bonus Section: Exercises to relieve stress at the computer and workstations

Order Form

Complete the following information and send this form with your check to **Letsdoyoga.com, 3360 McGraw Lane, Lafayette CA 94549**. (Please make check payable to Letsdoyoga.com) For credit card orders, order online at **www.letsdoyoga.com**

Name _____

Company _____

Street _____

City _____ **State** _____ **Zip** _____

Email _____ **Phone** _____

☐ Please send me the following items:

Item name _____

                              Quantity _____ **Price** _____

Item name _____

                              Quantity _____ **Price** _____

Item name _____

                              Quantity _____ **Price** _____

Item name _____

                              Quantity _____ **Price** _____

                                               Subtotal _____

                                               Sales tax _____

($3.50 for first item, $1.95 for each additional item.) **Shipping** _____

                                               Total _____

Order Form

Complete the following information and send this form with your check to **Letsdoyoga.com, 3360 McGraw Lane, Lafayette CA 94549**. (Please make check payable to Letsdoyoga.com) For credit card orders, order online at **www.letsdoyoga.com**

Name _____

Company _____

Street _____

City _____ **State** _____ **Zip** _____

Email _____ **Phone** _____

☐ Please send me the following items:

Item name _____

                          Quantity _____ **Price** _____

Item name _____

                          Quantity _____ **Price** _____

Item name _____

                          Quantity _____ **Price** _____

Item name _____

                          Quantity _____ **Price** _____

                                                  Subtotal _____

                                                  Sales tax _____

($3.50 for first item, $1.95 for each additional item.) **Shipping** _____

                                                        Total _____